苏槿 萧三闲 著／王国振 译
Writed by Sue Jane & Shane Shaw
Translated by Wang Guozhen

五洲传播出版社
China Intercontinental Press

二十四节气,是农历中表示季节变迁的24个特定节令,起源于两千多年前,至今仍然影响着中国人的生活和农牧业生产。二十四节气根据地球在黄道(即地球绕太阳公转的轨道)上的位置变化制定而成。每一个节气分别对应于地球在黄道上每运动15°所到达的一定位置,也就是说每个节气之间大约间隔15天。

二十四节气始于立春,终于大寒。立春、春分、立夏、夏至、立秋、秋分、立冬、冬至是二十四节气中最重要的八个节气,标志着季节转换,四季更替。

The 24 Solar Terms, which are the 24 Specific climate phenomena of four seasons, in the lunar calendar, originated more than 2,000 years ago and still affect Chinese people's life and farming and animal husbandry. The drafters of the calendar put in 24 solar terms (jie-qi) – approximately in a 15-day interval – to divide the lunar year, corresponding to the day on which the sun enters the first or 15 th degree of one of the 12 zodiacal signs.

The 24 Solar Terms start from "Beginning of Spring" and end with "Great Cold". Beginning of Spring, Spring Equinox, Beginning of Summer, Summer Solstice, Beginning of Autumn, Autumn Equinox, Beginning of Winter, Winter Solstice are the eight most characteristic Solar Terms. They identify clearly the climate changes of four seasons.

万物苏，山水俏；
风解冻，可探梅；
迎春花，案上吟。

Beginning of Spring,
The land turns a verdant green,
And the flowers are in full bloom.

/ 2月2～5日 / February 2-5 /

插新花
Arrange flowers

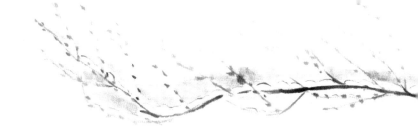

西 湖 早 春

The West Lake in
early spring

雨水
Rain Water

冬雪融，鸿雁来；
草木发，食春笋；
赏梅去，新茶俏。

The winter snow melts,
And the wild geese come back;
Grass and trees grow,
The plum blossoms look glorious,
And new tea is available.

／2月18～20日／ February 18-20 ／

雷声响，虫声鸣；
忙春耕，要吃梨；
正春光，玉兰开。

Thunder resounds and the insects sing joyfully;
Spring ploughing gets underway,
And the glorious magnolias bloom.

／3月5～7日／ March 5-7 ／

花入馔

Take cooking with flowers

子 规 啼

Cuckoo sings

Spring Equinox

春过半，花正浓；
放纸鸢，玩竖蛋；
雷为振，电鸣闪。

Halfway through Spring,
Flowers are in full bloom.
Paper kites fly exuberantly in blue sky.
Thunder resounds and lighting flash.

／3月20～22日／ March 20-22 ／

清明
Clear and Bright

宜踏青，宜祭祖；
摘蕨菜，食青团；
趋芳树，择园圃。

Go for outing and pay respects to departed ancestors.
Pick ferns and eat green rice balls.
Flowers Look glorious,
It is time to go for a walk in the park.

／4月4～6日／ April 4-6 ／

放 纸 鸢

Flying a kite

赏 牡 丹

To watch the peony

春已迟，春水涨；
蔷薇香，牡丹旺；
摘香椿，食鳜鱼。

Spring waters rise,
Roses and peony flourish;
Time to pick Chinese mahogany,
And catch mandarin fish.

/ 4月19～21日 / April 19-21 /

丁香开，草木香；
日初长，雷雨至；
蛙鸣叫，做蒲扇。

Cloves ripen; the daytime is longer;
Thunderstorms come,
Frogs sing and
It's a good time to make cattail fans.

／5月5～7日／ May 5-7 ／

树三鲜

Cherry, loquat, apricot

蕉下小酌

Drink under the
banana tree

小满
Grain Full

榴花燃，大河涨；
蚕结茧，稻谷满；
枇杷黄，苦菜香。

Pomegranates blossom.
Silkworms spin their intricate cocoons,
Rice grows sturdily,
And the loquats turn yellow.

／5月20～22日／May 20-22／

合欢开，稻谷肥；
江南雨，煮青梅；
小麦黄，蛙声鸣。

Albizia flowers blossom.
It's time to make plum wine.
Rice ears ripen, wheat turns yellow,
And frog choruses fill the night.

／6月5～7日／ June 5-7 ／

青 梅 酒

Plum wine

阳春面

Noodles in a simple sauce

骤雨急，折扇香；
一碗面，一口瓜；
等壶茶，观爱莲。

Sudden rain showers often occur,
Folding fans smell fragrant;
Melons look enticing, people sip tea,
And enjoy the blooming lotus in ponds.

/ 6月21～22日 / June 21-22 /

忙晒书，忙晒被；
做新饭，吃伏饺；
蝉嘶鸣，把暑叫。

People air books and quilts in the sunlight,
And cook new rice and summer jiaozi raviolis;
Cicadas begin their distinctive clatter.

／7月6～8日／ July 6-8 ／

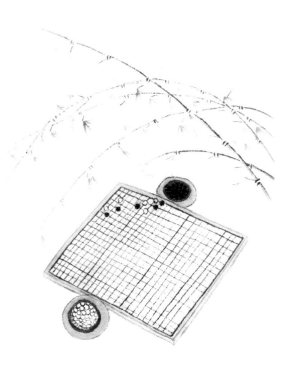

竹下对弈

Play the game of go

夏 收

The summer harvest

扑流萤，度盛夏；
羊肉汤，二伏面；
暑至极，候雷雨。

Great heat comes, glowworms fill the air at night,
Mutton broth and noodles are favored,
And thunderstorms come at any time.

/ 7月22～24日 / July 22-24 /

秋风起,是丰收;
牵牛花,稻谷香;
登高处,纳秋凉。

The autumnal harvests are celebrated,
The morning glory is enticing,
And the rice fragrance is delightful;
People climb up the mountain,
And enjoy the cool of autumn.

/ 8月7〜9日 / August 7-9 /

炖大肉

Braise meat

老鸭汤

The old duck boiled soup

处暑
Limit of Heat

放荷灯，祈福忙；
暑热止，秋风凉；
煲靓汤，喝茶凉。

Lotus lanterns offer a prayer for good luck,
Autumnal cool wind begin,
And people enjoy delicious broth,
And sip fragrant tea.

/ 8月22～24日 / August 22-24 /

芦花白，秋露美；
品秋茶，雁南归；
白露酒，白露茶。

Enticing white reeds and dew appear.
Autumn tea is favored,
Wild geese begin to head south,
And people drink white dew liquor and tea.

／9月7～9日／ September 7-9 ／

白露茶
White dew tea

吃肥蟹
Eat crab

枇杷花，谷入仓；
秋意高，天气凉；
烹肥蟹，品膏黄。

Loquat flowers blossom,
And grain fill the warehouses;
In the cooling weather of autumn,
People enjoy the fat crab.

/ 9月22～24日 / September 22-24 /

菊花黄，枫叶红；
秋意浓，忙收种；
登高远，思亲朋。

Yellow chrysanthemum and
Red maple leaves delight in high autumn days;
Farmers happily look on the bumper harvests.
Climb up the mountain and miss kinsfolk.

／10月8～9日／ October 8-9 ／

菊 花 酒
Chrysanthemum wine

霜 叶 红

Maple leaves

芙蓉傲，柿子软；
霜遍地，秋满园；
霜叶红，冬不远。

Hibiscus mutabilis look glorious and
Persimmons turn soft as frost is everywhere;
Maple tree leaves turn crimson and
Winter is at hand.

／10月23～24日／ October 8-9 ／

草木枯，冬渐浓；
庆秋收，忙冬种；
打边炉，快补冬。

Grass and trees wither,
Winter planting begins;
Hotpot is in vogue for warmth.

／11月7～8日／ November 7-8 ／

倭瓜馅饺子

Dumplings stuffed with squash

打边炉

Hot pot in Guang Dong

腊肉香，小雪飘；
杀年猪，吃刨汤；
熏腊肉，制香肠。

Eat delicious bacon as snow flakes dance;
People are busy butchering pigs to
Make bacon and sausages.

/ 11月22～23日 / November 22-23 /

大雪 / Great Snow

红梅开，天地白；
制腌肉，把年待；
冬进补，莫徘徊。

Red plums bloom,
And the land is whitened by snow;
People busily prepare salt meat for the new year,
And supplement their larders.

/ 12月6～8日 / December 6-8 /

冬白菜

Chinese cabbage

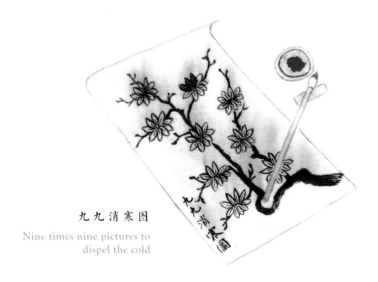

九九消寒图

Nine times nine pictures to dispel the cold

Winter Solstice

兰花暖,饺子香;
天地寒,冬夜长;
吃饺子,喝羊汤。

Orchids bloom and the weather is cold;
People eat jiaozi ravioli and
Mutton broth for extra calories.

/ 12月21～23日/ December 21-23 /

三九来,腊梅开;
粥端来,有热茶;
到年下,围炉坐。

Watersweet blooms in early winter,
Hot porridge and hot tea are served to keep warm.
Approaching the Spring Festival,
People sit around the burning stove to chat.

／1月5～7日／ January 5-7 ／

吃菜饭

Vegetable rice

岁朝清供图

Flowers and fruits of the
wayside and woodland

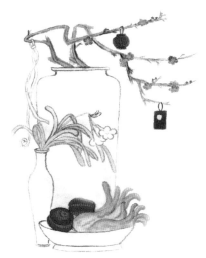

水仙开，人团圆；
吃年糕，备年饭；
贴窗花，过大年。

The narcissus blooms and people make and
Eat glutinous rice cakes,
And gather for the Spring Festival Eve dinner.
House windows are pasted with flower papercuts.

/ 1月20～21日 / January 20-21 /

图书在版编目（CIP）数据

四时：汉英对照 / 苏槿，萧三闲著 . -- 北京：五洲传播出版社，2019.6

ISBN 978-7-5085-4188-4

Ⅰ.①四… Ⅱ.①苏… ②萧… Ⅲ.①二十四节气 - 通俗读物 - 汉、英 Ⅳ.① P462-49

中国版本图书馆 CIP 数据核字 (2019) 第 087361 号

作　　者	苏　槿　萧三闲
翻　　译	王国振
出 版 人	荆孝敏
责任编辑	梁　媛
装帧设计	红方众文　优　昙　朱丽娜
出版发行	五洲传播出版社
地　　址	北京市海淀区北三环中路 31 号生产力大楼 B 座 6 层
邮　　编	100088
发行电话	010-82005927，010-82007837
网　　址	http://www.cicc.org.cn，http://www.thatsbooks.com
印　　刷	天津图文方嘉印刷有限公司
版　　次	2019 年 6 月第 1 版第 1 次印刷
开　　本	787mm×1092mm　1/32
印　　张	5
字　　数	20 千
定　　价	58.00 元